Pushes and Pulls

In My House

Ryan Smith

LIGHTBOX
openlightbox.com

LIGHTBOX

Go to **www.openlightbox.com** and enter this book's unique code.

ACCESS CODE

LBXZ3738

Lightbox is an all-inclusive digital solution for the teaching and learning of curriculum topics in an original, groundbreaking way. Lightbox is based on National Curriculum Standards.

OPTIMIZED FOR

- ✓ TABLETS
- ✓ WHITEBOARDS
- ✓ COMPUTERS
- ✓ AND MUCH MORE!

STANDARD FEATURES OF LIGHTBOX

AUDIO High-quality narration using text-to-speech system

VIDEOS Embedded high-definition video clips

ACTIVITIES Printable PDFs that can be emailed and graded

WEBLINKS Curated links to external, child-safe resources

SLIDESHOWS Pictorial overviews of key concepts

INTERACTIVE MAPS Interactive maps and aerial satellite imagery

QUIZZES Ten multiple choice questions that are automatically graded and emailed for teacher assessment

KEY WORDS Matching key concepts to their definitions

VIDEOS

WEBLINKS

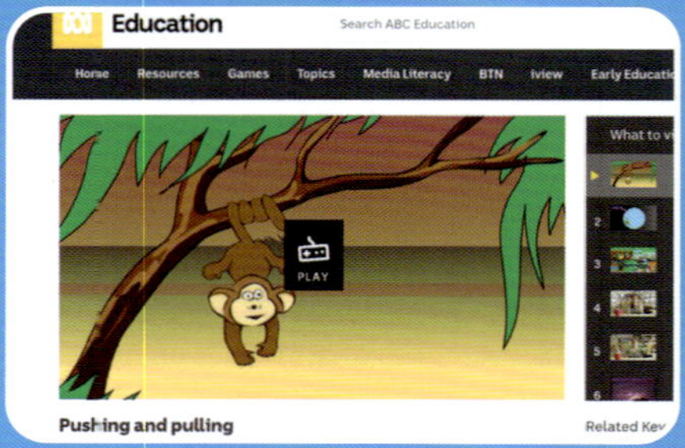

SLIDESHOWS

QUIZZES

Pushes and Pulls

In My House

Contents

We push and pull things every day. We push things to move them away from us. We pull things to move them closer.

My family and I are cleaning our house today.

We can learn about the science of pushes and pulls while we clean.

My brother pushes the vacuum over the floor.

The floor is easy to vacuum.

He vacuums the rug next. It is harder to push the vacuum there.

Why is it harder to push the vacuum on the rug?

The **first vacuums** were so **big** **two people** were needed to use them.

I make my bed.

I pull my blanket across the bed and smooth it out.

More of my blanket is hanging off one side of my bed.

How can I make my blanket even on both sides?

We are going to do laundry next.

I push my empty laundry basket.
It is easy to push.

I fill my basket
with dirty clothes.
It is harder
to push the
basket now.

Why is the basket harder to push?

I lost a toy
behind a chair.

My sister pulls the chair forward.
Now I can grab my toy.

My brother sits down in the chair. It is harder for my sister to push the chair back.

Why is it harder to push the chair back?

An armchair can weigh as much as one small bale of hay.

See what you have learned about pushes and pulls in your house.

What do you push in your house?
What do you pull in your house?

KEY WORDS

Research has shown that as much as 65 percent of all written material published in English is made up of 300 words. These 300 words cannot be taught using pictures or learned by sounding them out. They must be recognized by sight. This book contains 60 common sight words to help young readers improve their reading fluency and comprehension. This book also teaches young readers several important content words, such as proper nouns. These words are paired with pictures to aid in learning and improve understanding.

Page	Sight Words First Appearance
4	and, away, day, every, from, move, them, things, to, us, we
5	about, are, can, family, house, I, learn, my, of, our, the, while
6	over
7	is
8	he, it, next, on, there, why
9	big, first, people, so, two, use, were
10	make
11	out
12	more, off, one, side
13	both, even, how
14	do
15	in, world
16	now, with
18	a
20	back, down, for
21	an, as, much, small

Page	Content Words First Appearance
5	pulls, pushes, science
6	brother, floor, vacuum
8	rug
10	bed
11	blanket
14	laundry
15	basket, Illinois, laundromat
16	clothes
18	chair, toy
19	sister
21	armchair, bale, hay

Published by Smartbook Media Inc.
14 Penn Plaza, 9th Floor New York, NY 10122
Website: www.openlightbox.com

Library of Congress Control Number: 2020014794

ISBN 978-1-5105-5462-7 (hardcover)
ISBN 978-1-5105-5463-4 (multi-user eBook)

Printed in Guangzhou, China
1 2 3 4 5 6 7 8 9 0 24 23 22 21 20

052020
110819

Designer: Ana María Vidal Project Coordinator: Ryan Smith

The publishers acknowledges Getty Images, iStock, and Shutterstock as the primary image suppliers for this title.